FLOWER GARDENS

START TO PLANT

FLOWER GARDENS

*Create your ideal garden with these
simple-to-follow projects*

GRAHAM A PAVEY

GREENWICH EDITIONS

A QUANTUM BOOK

Published by Greenwich Editions
10 Blenheim Court
Brewery Road
London N7 9NT

ISBN 0-86288-143-9

QUMHFG

This book was produced by
Quantum Books Ltd
6 Blundell Street
London N7 9BH.

Manufactured in Singapore by Eray Scan Ltd.
Printed in Singapore by Star Standard Industries Pte
Ltd.

AUTHOR'S ACKNOWLEDGMENTS

I would like to give special thanks to my wife, Chris; to Steve Woods and
the staff at Tacchi's Garden Scene, Wyton, Huntingdon, Cambs, PE17 2AA;
and to Ian and Susie Pasley-Tyler, of Cotonn Manor, near Guilsborough,
Northants.

CONTENTS

INTRODUCTION

For a stunning border that brightens up the garden from season to season and requires a low level of maintenance, planning is the key. By choosing plants carefully to ensure that they suit the aspect, the soil and are suitable companions for their neighbours, you can be sure of a changing display where there will nearly always be some attractive perennial to admire.

Constantly changing, the mixed border offers an annual tapestry of colours. Bulbs may dominate in spring; lush, majestic flowers appear in summer; until the warm and rich colours of autumn arrive with a flourish. In winter, a solid structure of evergreens ensures all-year-round interest.

A modestly sized border can contain a mass of interesting plants. It should be located to ensure that it can be enjoyed during its flowering season. A winter or spring border should be visible from the house so that you can get as much pleasure from it as possible. All you need is a sunny situation and a well-nourished soil that is not too heavy. Each project in this book offers a professionally designed border to the beginner, removing the doubt and ensuring a successful display.

MATERIALS AND TECHNIQUES

Designing the garden

Before looking at individual border projects, some thought should be given to the overall garden and how simple design ideas can make a dramatic difference. The layout illustrated here is for a small and basic garden, but similar ideas could be used in any size of garden.

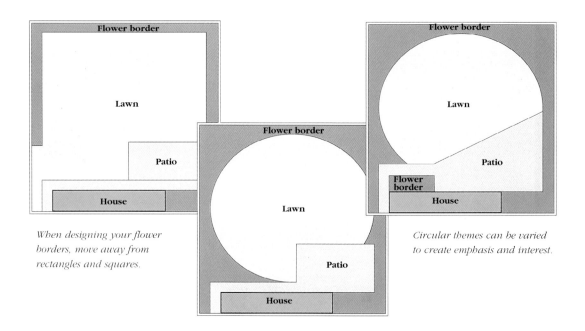

When designing your flower borders, move away from rectangles and squares.

Circular themes can be varied to create emphasis and interest.

LAYOUT 1

Many new gardens have this kind of layout, with the house at one end and the garden behind it, often surrounded by a solid, wooden fence. There may be a small patio close to the house and a rectangular lawn with a narrow border, around the perimeter of which unprepared soil is planted with whatever plants looked good at the garden centre on the day of purchase.

LAYOUT 2

First, try to forget about the rectangles and squares created by the site, and think of other shapes. Start by overlaying the garden with a circle. A circular lawn, for instance, will create large beds in each corner – borders with room to work in. The projects in this book could be used in any one of the corner beds created here.

LAYOUT 3

By changing the angle of the patio, we can alter the whole emphasis of the garden and push the sitting area out from the house, to pick up more of the sun in north-, east- and west-facing sites. The circular theme can now be 'hung' off the patio and varied to suit the site.

Preparing the ground

One of the keys to success is in the preparation of the soil – the more thoroughly the soil is improved, the better the final result. This preparation starts with digging in plenty of organic matter (see page 14, 'Improving the soil'), which generally comes in the form of compost or well-rotted animal manure and sharp sand to lighten it and improve drainage. In a temperate climate where the soil has a clay content, the preparation is best carried out in the autumn and the soil left unplanted over the winter. This will allow the frost to work on it and break it down further.

1. Dig out a trench across the whole width of the bed to one spade's depth, and place this soil in a wheelbarrow, or to one side.

2. Line the base of the excavation with a 4 in-/10 cm-deep layer of organic matter, in this case well-rotted horse manure.

3. Dig out a second trench alongside the first.

4. Turn this soil onto the organic matter in the first trench.

5. Line the new trench with organic matter and repeat the cycle across the bed. Once the bed has been completely dug over, fill the final trench with the soil removed from the original one.

6. Once the ground has been prepared, or in the following spring, rake it to create a fine tilth ready for planting.

Soil testing

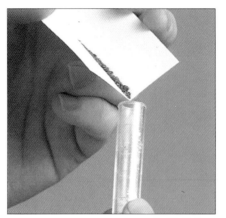

1. Place testing powder in the test tube to the first mark and then add the soil to the second mark.

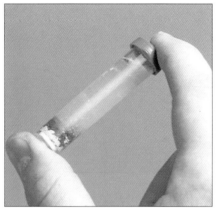

2. Fill the tube with *tap* water (the powder is balanced to accept alkaline tap water) and shake thoroughly.

3. Leave the test tube to stand until the residue settles. With some soils this may take a long time.

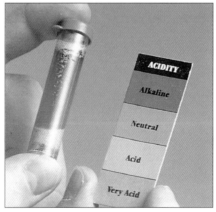

4. Compare the colour of the solution with the accompanying colour chart to determine the pH. This soil is alkaline.

Before deciding what preparation the soil needs or what plants can be planted, the soil needs analysing.

Acid or alkaline

The first stage of this is to determine the pH of the soil, which indicates whether the soil is acid or alkaline. There are some plants, for instance, such as rhododendrons, most heathers and azaleas, that need an acid soil and cannot grow in any other. Testing the pH of the soil can easily be done using a small kit available at any garden centre. Follow the instructions on the test kit.

Soil structure

Once you have determined the pH of your soil, then look at the soil structure. Take a walnut-sized piece of soil, wet it and attempt to roll it into a ball between the palms of your hands. If it keeps falling open with no adhesion at all, then the soil is a sandy one; if it can be formed into a smooth ball, then it is very heavy clay. Any form of ball indicates a clay content.

Sandy soils are open, free-draining and low in nutrients, as these are continually being washed from the soil. Improve this type of soil by digging in copious amounts of well-rotted organic matter, which acts as a sponge, holding water and nutrients.

Clayey soils are difficult to work, but higher in nutrients. Improve clay soil by digging in organic material mixed with some form of grit to improve drainage.

Improving the soil

As well as well-rotted organic matter, there are other materials which can be incorporated into the soil, depending upon the conditions. Mix them in equal proportions with the well-rotted manure in Steps 1 to 5 on pages 11 to 12, 'Preparing the ground'.

Clayey soil can make life difficult in the garden. Being very claggy in winter and rock hard in summer, it is difficult to work, having very tiny particles which cling tightly together. As sand has large particles which are very open and do not cling together, the solution is to incorporate sand into the clay soil to open it up. The best sand for this is sharp sand, but any gritty material would be suitable, as long as it is clean.

Peat is a natural material obtained from peat bogs and is excellent for conditioning the soil. Mixed in equal parts with manure or compost, it produces the very best material for preparing garden soil. Its removal has a detrimental impact on the natural environment and it is recommended that if peat is to be used only that extracted from managed bogs should be used.

Coco-fibre or **coir** is a material that has been heralded as the revolutionary new alternative to peat, although it was in common use in the nineteenth century. From the environmental angle there is a question mark over it because it is said to be impoverishing the soil on the islands in which it is found. Some environmentalists have now returned to using peat from managed sources. Use it in the same way as peat.

Loam is soil of medium texture that contains more or less equal parts of sand, silt and clay and is usually rich in humus. Where possible, improving the existing soil is always the best solution, but in some instances, where the soil is poor, it is necessary to bring in a good-quality loam to replace the soil which already exists. Always ask for 'screened' topsoil, as this has been treated to destroy any weeds or weed seeds. Once the new soil is in place it would still be best to dig the border as described in 'Preparing the ground'.

Mulching

Once the border has been planted, the secret of success is to keep the roots moist and cool throughout the summer. This can best be achieved by covering the soil around the roots when wet, in the spring, with a mulch. It can also help to keep weeds down by preventing seeds from germinating and perennial weeds from growing by eliminating light.

Almost anything can be used as a mulch, from newspaper and old carpet to grass cuttings, but as the flower border is to be aesthetically pleasing, materials such as forest bark, pea shingle or mulching mats are especially recommended.

Forest bark is waste material from the logging and timber industry. It comes in several different grades and is also used for soil conditioning and as a soft paving material in children's play areas. It must be 2 in/5 cm deep to be effective. Coco-fibre could be used as a viable alternative.

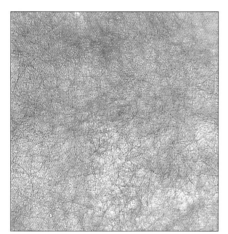

Mulch matting is an interwoven material usually made of polypropylene, fibreglass or a similar man-made material and comes in a roll or as small squares or circles for individual plants. The weave allows water through to the soil while still keeping weeds at bay.

Pea shingle is a very effective mulch, but immediately alters the character of the border and may influence the overall design of the garden, so care must be used in its selection.

Planting

Plants, like all living things, respond to thoughtful treatment. Care at planting time can pay dividends in the long run.

Dig a hole with a garden trowel sufficient to take the whole rootball. Before planting, put the plant, in its pot, into the hole to ensure that the top of the soil in the pot is flush with the top of the surrounding soil. For the holes for larger plants, use a spade.

When you remove the plant from its pot to plant it you may find that it has become 'root-bound' in its pot and, if planted in this state, any new roots will not invade the surrounding soil. It is therefore a good idea gently to 'tease' out some of the roots before planting.

Once planted, consolidate the soil around the roots by gently firming it down with your fingertips. With larger plants, use the heel of your shoe.

Plant sizes

The size of the plant at the time of planting can be critical. If it is too small it could be swamped by its neighbours before it has time to grow; too large and it may initially be out of proportion with its neighbours. In the projects that follow the ideal pot size for each plant is indicated, but if these are difficult to obtain then plants in larger pots would be acceptable.

Watering

Once each plant has been planted, the roots should be given a good drenching from either a watering can or a gentle trickle from an open-ended hose pipe. This will remove any air pockets from around the roots, as well as give the plant the water it needs for a good start.

Plant protection

Some plants will require protection during a cold winter and this is best done by wrapping them up in old net curtains or a sheet of horticultural fibre, a 'fleece' usually made of woven polypropylene.

Supporting climbers

Most climbers will need some support to hold them against the structure chosen to accommodate them. Various methods may be employed, from plastic netting to a solid, wooden trellis, but the problem here is that the support is very visible and spoils the effect of the climber. The best solution is to use vine eyes and galvanised wire, which blend with the background and are virtually invisible. Attach three eyes, vertically, equally spaced, to each fence post, or about 6 ft/ 1.8 m apart on a wall.

but regular watering, especially in dry weather, will be essential to success. There are a number of excellent fine-spray attachments now available which gently water the plants without causing damage.

A **watering can** could be used for individual plants when planting up and for the close watering of individual plants that may need extra moisture.

A **wheelbarrow** is essential when preparing the ground for ferrying manure, soil, grit or other materials to where they are needed.

Galvanised wire is used with vine eyes for 'wiring' a wall or fence to provide support for climbers.

Vine eyes are either screws, ideal for walls or screwing straight into a fence post, or masonry nails, which could also be used in a fence post. Galvanised wire is fastened to vine eyes to provide a network of wires to support climbing plants.

Tools

A **spade** is an essential tool for preparing the border, either for edging or shovelling preparation materials. A smaller version, called a border spade, can be used for digging around existing plants without damage.

Once the garden has been dug over, a **rake** can be used to smooth the surface, creating a fine tilth. Draw it backwards and forwards across the bed in different directions, breaking up any large lumps with a sharp slap, until the desired surface has been created.

A **gardener's trowel** is a useful small tool for digging small holes ready for planting.

A **hose** is needed once the border has been planted. Initially, watering helps to bed the plants in and remove any air pockets which may have developed while planting,

Screw eyes provide support that blends into the background.

EAST-FACING CORNER BED

Dry, with only early morning direct sunlight, the east-facing border can cause problems, so plants which can cope with these conditions have been selected. For most of the year this scheme relies on a variety of leaf shapes and colours, but from mid-summer through to autumn the border will come to life with the Crocosmia, *which is followed by autumn-leaf colours.*

The plants

Astrantia major, 3½ in/9 cm or 5½ in/
1 litre pot

2 Crocosmia 'Lucifer', 3½ in/9 cm or
5½ in/1 litre pots

2 Hebe rakaiensis, 7½ in/3 litre pots

3 Hosta sieboldiana 'Elegans', 5½ in/1 litre or
6½ in/2 litre pots

6 Houttuynia cordata 'Chameleon',
3½ in/9 cm or 5½ in/1 litre pots

Lonicera japonica 'Halliana',
7½ in/3 litre pot

Prunus lusitanica, 7½ in/3 litre pot

Vitis coignetiae, 7½ in/3 litre pot

Tools

Spade • Garden trowel

Quick tips

Astrantia major.

Flowering season: the main season will
be from mid-summer to early autumn.
Soil: any good garden soil.
Care: the laurel (Prunus lusitanica) may
require an annual prune to restrain it. Slug
pellets placed around the hostas in the
spring will deter snails and slugs from
feasting on their leaves, although H.
sieboldiana appears to be less palatable to
them than other hostas. The climbers will
need tying to the wires.

Crocosmia 'Lucifer'.

Planting plan

Main structural planting

The first plants to consider are the main ever-greens and structural plants. Evergreens at each corner give the scheme balance and ensure all-year-round interest.

Prunus lusitanica (Portuguese laurel). This reliable evergreen will supply the main structure for this grouping. It has the advantage of growing anywhere, sun or shade, and flowers in early summer. It is a large plant, eventually forming a small tree, but can be pruned annually, once established, to keep it in check.

Vitis coignetiae (Japanese crimson-glory vine). This vine has large, round, deciduous leaves which turn crimson in the autumn, and is always an eye-catcher. The trunk becomes old and gnarled in time, giving a solid structure despite the lack of leaves in the winter.

Hebe rakaiensis. Large-leafed hebes are, unfortunately, not reliably hardy in temperate zones, but those with smaller leaves are quite tough. This hebe has small leaves and will form a small, solid mound covered in white flowers during mid-summer. It is the perfect plant for the corner of a bed, where it quickly establishes itself as a reliable structural plant.

***Lonicera japonica* 'Halliana'** (honeysuckle). This evergreen climber will reliably smother any wall or fence as long as the structure is carefully 'wired'. It is not as fragrant as other honeysuckles, but compensates for that by providing a solid, evergreen screen.

Astrantia major (masterwort). With its attractive, feathery foliage, which lasts all summer, and its rather curious cream-and-green flowers from mid- to late summer, masterwort contrasts well with more solid shapes. Here it is used as a foil for the hebe and the large leaves of the hosta and the vine.

STEP TWO
In-fill planting

In a small bed or garden, it is useful to choose plants which offer 'value for money', providing interest over a long period with attractive foliage, as well as flowers.

Hosta sieboldiana 'Elegans' (plantain lily). Hostas are excellent plants for associating with others. Their large, often variegated, leaves provide a foil for sword-shaped and feathery foliage. Most require shade and some moisture in the soil, but the *H. sieboldiana* varieties will cope with a dry soil and will even grow in full sun.

Houttuynia cordata 'Chameleon'. This marsh plant may spread freely through moist soil, becoming very invasive. It is one of the few plants which will grow in both boggy and dry conditions, and in dry soil it is much more restrained. The unusual, aromatic, pink--, cream- and green-variegated leaves offer interest throughout the summer, its white flowers being a bonus.

Crocosmia 'Lucifer' (montbretia). The sword-shaped leaves appear early in the season and contrast well with so many other plants. In late summer, the *Crocosmia* adds its flame-red flowers to the scheme, where the *Astrantia* will still be flowering. Combining with the *Houttuynia*, it completely changes the character of the arrangement to become hot and vibrant, giving it a lift at a time when other plants are beginning to look a little tired.

SOUTH-FACING CORNER BED

This aspect, in full sun for most of the day, gives the opportunity to grow sun-loving plants which may be difficult to grow elsewhere. Not all plants need full sun, so care must be taken to select plants which enjoy these hot, dry conditions. A corner bed of this nature, facing south and backed by a wall or fence, could be planted up with bold, hot colours to take advantage of the condition.

The plants

Artemisia 'Powis Castle', 7½ in/3 litre pot

Convolvulus cneorum 7½ in/3 litre pot

Cotinus 'Grace', 7½ in/5 litre pot

2 *Geranium psilostemon*, 3½ in/9 cm or 5½ in/1 litre pots

3 *Osteospermum jacundum*, 5½ in/1 litre or 6½ in/2 litre pots

4 *Rosa* 'Little White Pet', 7½ in/3 litre pots or bare-rooted

3 *Sisyrinchium striatum*, 3½ in/9 cm or 5½ in/1 litre pots

2 *Stachys olympica* 'Silver Carpet', 3½ in/9 cm or 5½ in/1 litre pots

2 *Trachelospermum jasminoides*, 7½ in/3 litre pots

Tools and materials

Galvanised wire • Garden trowel • Spade • Vine eyes

Quick tips

Geranium psilostemon.

Flowering season: mid-summer.
Soil: any garden soil. Heavy soil should be broken up with the addition of sharp sand.
Care: the *Trachelospermum* should be tied to the wires regularly. They must be protected in cold weather, perhaps using horticultural-fibre fleece, as severe wind-chill and frost will cut this plant hard back and may kill young specimens. Prune the *Cotinus* down hard each spring to encourage larger and more colourful foliage. The leaves of *Sisyrinchium* blacken as they die, which is perfectly normal, but should be removed as soon as they appear. Cut the *Artemisia* to within 4 in/10 cm of the ground each spring to maintain its shape.

Planting plan

Main structural planting

Cotinus 'Grace' (smoke tree). The large, purple leaves of this deciduous shrub dictate the overall colour scheme of whatever border it is planted in. Despite being deciduous, the bare branches have a solid shape, contributing to the overall framework.

Trachelospermum jasminoides. Like an evergreen jasmine, this plant is the perfect choice for a sheltered wall. It must be protected in very cold weather (see 'Care').

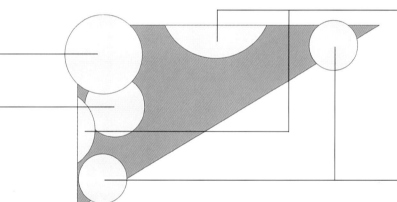

Artemisia 'Powis Castle'. This is a foliage plant which is perfect for a myriad of uses. It is equally at home with roses in a cottage garden as it is with box hedging in a more formal setting. It is important to cut the plant hard back every spring to maintain its shape.

Convolvulus cneorum. This silvery, glistening relative of bindweed needs careful placing in the garden, as it must have full sun or it will become untidy and straggly.

STEP TWO
In-fill planting

Geranium psilostemon. This is the tallest-growing of all hardy geraniums. When its bright, magenta flowers appear in mid-summer, it is like welcoming an old friend who only visits once a year. The geraniums seem to change colour when viewed with the purple leaves of the *Cotinus* in the background, giving them an almost luminescent quality.

***Stanchys olympica* 'Silver Carpet'** (lambs' tongues). This is a non-flowering variety, grown for its evergreen, mat-forming, woolly, grey leaves.

Sisyrinchium striatum. One of the joys of summer. The sword-shaped leaves and upright flower stems contribute to the upright, spiky display, and creamy-yellow flowers appear in mid-summer.

Osteospermum jacundum. White, daisy-like flowers cover this plant from early summer through to autumn, an ideal way of extending the flowering season in a small border.

***Rosa* 'Little White Pet'**. This patio rose also makes good ground cover if left alone. The white, pompom flowers are produced from middle to late summer.

WEST-FACING CORNER BED

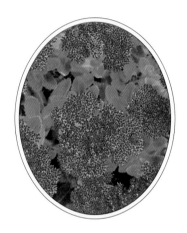

This aspect means that the bed will not be in full sun until the afternoon.
The sun will have reached its hottest by this time, so the area will warm up
very quickly, making it a good choice for plants which require a sheltered site.
Here you can grow cottage-garden plants that require little maintenance and
which may not grow in other parts of the garden.
A predominantly pink scheme has been chosen, which is a good
colour for a west-facing border, and it will be highly scented in the
early to middle part of the summer.

The plants

3 *Agapanthus* 'Loch Hope',
5½ in/1 litre pots or larger

Artemisia 'Powis Castle', 7½ in/3 litre pot

Itea ilicifolia, 7½ in/3 litre pot or larger

Jasminum officinale,
7½ in/3 litre pot or larger

3 *Rosa* 'President de Seze', 7½ in/3 litre pots
or bare-rooted

2 *Salvia officinalis*, 6½ in/2 litre or
5½ in/1 litre pots

3 *Sedum spectabile*, 3½ in/9 cm or
5½ in/1 litre pots

Solanum crispum 'Glasnevin's Variety',
7½ in/3 litre pot or larger

Tools and materials

Garden trowel • Spade • Galvanised wire
• Vine eyes

Quick tips

Agapanthus 'Loch Hope'.

Flowering season: early summer through to early autumn.
Soil: any good garden soil. Improve the soil each spring with well-rotted organic matter.
Care: these roses do not need pruning, but any dead wood should be removed in early spring. Cut the *Artemisia* down hard in mid-spring and also the sage if it has become poor and leggy in the winter. The climbers should be tied into wires on the wall or fence. *Agapanthus* is an excellent subject for a container, where it flowers better because its space is restricted. It must be brought into frost-free conditions to over-winter in very cold areas.

Planting plan

Main structural planting

***Solanum crispum* 'Glasnevin's Variety'** (climbing potato). The rich, blue flowers of this climber, carried over a long period in mid-summer, are good companions for a variety of colour schemes.

Itea ilicifolia (sweetspire). This evergreen needs a protected site, and its loose habit means that it is best grown against a wall or fence. The hanging flowers are sweetly scented and appear in late summer.

***Rosa* 'President de Seze'.** The Gallica rose is probably the oldest of all garden roses, having been grown by the Greeks and Romans. Although only flowering once in early to mid-summer, this Gallica rose provides a magnificent display, not to be missed. Attractive buds open to a full flower with a subtle mix of purple, violet, brown and grey, with a sweet scent. Here the *Artemisia,* sage and rose provide a perfect silver-and-pink combination.

Jasminum officinale (jasmine). This summer-flowering jasmine is quite accommodating and will even grow on a shady wall. Sweetly scented, it is an essential ingredient for a pretty, country-garden border.

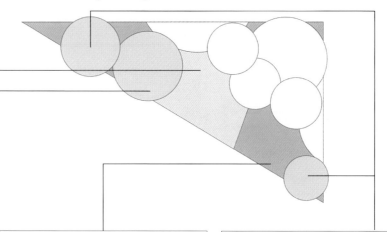

***Agapanthus* 'Loch Hope'**. One of the most spectacular of all late-summer-flowering plants. Its blue flowers, growing in a candelabra style on the end of a long, sturdy stem, are unforgettable and will contrast with the pinks of the scheme to change the overall character. In early summer its strap-like leaves will provide a contrast in shape and texture. Later in the season the emphasis will shift to the *Agapanthus*-and-*Sedum* combination, which takes up the baton for the end of the summer.

***Artemisia* 'Powis Castle'.** A stately plant, which always draws attention in the summer garden and has many uses in planting design. The grey foliage tones down any bright colours and helps to link shades which might otherwise be incompatible. It is a good choice in a traditional border, combining well with both old roses and herbaceous plants.

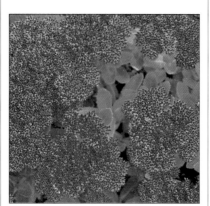

Sedum spectabile. Grey leaves in early summer combine well with the greys and pinks of the arrangement, and the pink flowers, much loved by butterflies, flower with the *Agapanthus* in late summer to continue the flowering season.

Salvia officinalis (sage). Matte, grey-green, evergreen leaves and a good shape combine to create a border plant of some distinction. It is the perfect foil for old roses and other country-garden plants. Its understated presence does not interrfere with flowering neighbours.

NORTH-FACING CORNER BED

This is the shadiest of all the corner beds, but it is open to the sky
and should therefore not be regarded as being in full shade.
Many plants prefer this aspect to a south-facing one, so the choice
is not too limiting. Generally, colourful plants are not happy in
shade, so the best approach is to create a scheme relying on shape
and subdued colours. This scheme uses shades of green,
some white, a little yellow and one bold splash of red
in the leaves of the Heuchera.

31

The plants

2 *Alchemilla mollis*, 3½ in/9 cm
or 5½ in/1 litre pots

2 *Bergenia cordifolia*, 3½ in/9 cm
or 5½ in/1 litre pots

Elaeagnus x *ebbingei* 'Limelight',
7½ in/3 litre pot

4 *Epimedium* x *versicolor*
'Neo-Sulphureum', 3½ in/9 cm
or 5½ in/1 litre pots

3 *Heuchera* 'Purple Palace', 3½ in/9 cm or
5½ in/1 litre pots

2 *Hydrangea petiolaris*, 7½ in/3 litre pots

Iris pallida 'Variegata',
3½ in/9 cm or 5½ in/1 litre pot

2 *Rosa* 'Iceberg', 7½ in/3 litre pots
or bare-rooted

Tools

Garden trowel • Spade

Quick tips

Rosa 'Iceberg'.

Flowering season: the flowers are less important in this arrangement than in others, but there will be blooming from spring into mid-summer.

Soil: any good garden soil. Improve the soil each spring with well-rotted organic matter.

Care: remove any dead wood from the roses in early spring and any dead material, such as leaves, from the perennials as necessary. The climbing hydrangea is renowned for looking very sickly for three or four years after planting. This is nothing to worry about as the plant is concentrating on creating a sound root system, after which it will quickly cover the wall or fence.

Planting plan

Main structural planting

Hydrangea petiolaris (climbing hydrangea). This spectacular plant will grow in quite deep shade and has the added advantage of being self-clinging. Although not evergreen, its woody trunk and branches, with shaggy bark, make a solid structure in the winter.

Elaeagnus x ebbingei 'Limelight'. This reliable, fast-growing evergreen brightens the garden at any time of year with its subtle, silver-gold-variegated leaves.

Epimedium x versicolor 'Neo-Sulphureum' (motherwort). *Epimedia* make excellent, evergreen, ground-cover plants for shady areas, especially in woodland. The pale-yellow flowers appear in early spring, but are often hidden under the foliage. The top foliage can be cut away to reveal them, if desired.

Heuchera 'Purple Palace'. There are many varieties of this evergreen plant, but this is probably the best and most commonly grown. It has larger leaves than other varieties and the purple leaves can be used either as a contrast to yellows or a companion for reds and pinks.

In-fill planting

***Rosa* 'Iceberg'.** Many roses are very useful shrubs for a mixed planting scheme and 'Iceberg' is no exception. Flowering for most of the summer, the small leaves and white flowers of *Rosa* 'Iceberg' make a startling combination and will work well in either sun or shade.

***Iris pallida* 'Variegata'.** This attractive, variegated iris adds a dramatic effect to the overall scheme, contrasting with the feathery foliage of both the lady's mantle and the fern. The sword-shaped leaves of irises are a good way of adding interest to an otherwise 'flat' planting scheme.

Bergenia cordifolia (elephant's ears). *Bergenia* are the perfect companion for many plants. Here they combine well with *Heuchera* 'Purple Palace', making a composition which works throughout the year. In small groups dotted around the garden, they help to tie a scheme together and enhance the other plants, although *en masse Bergenia* can look untidy and overpowering.

Alchemilla mollis (lady's mantle). Growing in sun or shade, the lime-green, rounded, crinkly-edged foliage of this plant has no equal.

EAST-FACING LONG BORDER

These long borders are designed for medium to large gardens. For longer borders, duplicate along the whole length to ensure that the border is a balanced success. This east-facing border will look its best early in the day, when the yellows and blues will be lit up by the sun. A good position for it would be close to where breakfast is taken, so that it can be viewed either from a seating area in the garden or an appropriate window.

The plants

Aucaba japonica 'Variegata', 7½ in/3 litre pot

2 *Buxus sempervirens* 'Elegantissima', 7½ in/3 litre pots

Clematis macropetala, 6½ in/2 litre or 7½ in/3 litre pot

10 *Delphinium* 'Galahad', 3½ in/9 cm or 5½ in/1 litre pots

Euonymus fortunei 'Silver Queen', 6½ in/2 litre or 7½ in/3 litre pot

5 *Geranium* x *riversleaianum* 'Russell Prichard', 5½ in/1 litre or 6½ in/2 litre pots

5 *Hosta sieboldiana* 'Frances Williams', 5½ in/1 litre or 6½ in/2 litre pots

5 *Iris germanica*, 3½ in/9 cm or 5½ in/1 litre pots

2 *Mabonia japonica*, 7½ in/3 litre pots

2 *Rosa* 'Graham Thomas' (Registered name: 'Ausmas'), 7½ in/3 litre pots or bare-rooted

2 *Parthenocissus tricuspidata* 'Veitchii', 7½ in/3 litre pots

Philadelphus 'Sybille', 7½ in/3 litre pot

2 *Pleioblastus viridistriatus*, 7½ in/3 litre pots or larger

3 *Thalictrum aquilegiifolium*, 3½ in/9 cm or 5½ in/1 litre pots

3 *Viburnum opulus* 'Compactum', 6½ in/ 2 litre or 7½ in/3 litre pots

Quick tips

Viburnum opulus.

Flowering season: there are odd splashes of colour throughout the year, but the main flowering season will be in late spring through to mid-summer. A combination of leaf shapes, shades of green and different textures will maintain interest for a longer period than using flower colour alone.

Soil: any good garden soil. Improve the soil each spring with well-rotted organic matter.
Care: stake the delphiniums with strong canes in spring. Cut any dead wood out of the roses in early spring.

Tools

Garden trowel • Spade

Planting plan

STEP ONE
Main structural planting

Mahonia japonica. This architectural plant looks good against a plain background where the foliage can stand out. Here its evergreen foliage contributes to the structure of the arrangement. Scented flowers will appear in mid-winter and some may be out at Christmas.

***Buxus sempervirens* 'Elegantissima'** (variegated box). Box is an excellent structural plant for use in either a formal setting or an informal, cottage-garden style. The plain-green plant is a large growing shrub, eventually turning into a small tree, and needs constant attention. This variegated form is much better behaved and needs less effort.

***Philadelphus* 'Sybille'** (mock orange). Spectacular, bell-shaped flowers cascading down arching branches, combined with a sweet, orange scent, make this plant one of the most popular for mid-summer flowering.

STEP TWO
Main structural planting

Aucuba japonica **'Variegata'** (spotted laurel). This is a truly excellent plant, growing in any condition, including very deep shade. In spring its golden foliage seems to glow when other plants are looking tired after the winter months.

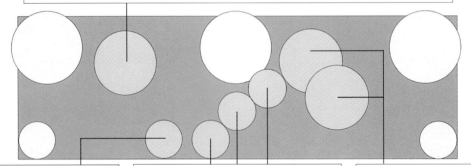

***Euonymus fortunei* 'Silver Queen'**. There are a large number of different varieties of *Euonymus fortunei* – all of them easy, reliable evergreens. They can all be clipped like a hedge and many will even climb a wall if grown against one.

***Viburnum opulus* 'Compactum'.** This dwarf guelder rose is a much better plant than its larger cousin. Its deciduous foliage comes right down to the ground, smothering any weeds, and the feathery foliage is an ideal companion for more solid leaves. Its white flowers appear in early summer.

Pleioblastus viridistriartus. This bright-yellow bamboo grows only to about 4 ft/1.2 m high and, unlike some other bamboos, is quite well behaved. Here, its upright growth contrasts with the architectural shape of its neighbours.

STEP THREE
In-fill planting

Parthenocissus tricuspidata **'Veitchii'**
(Boston ivy). Related to the Virginia creeper, this
is a much better plant. It will quickly grow over
a wall or fence, whereas the Virginia creeper
would rather scramble along the ground.

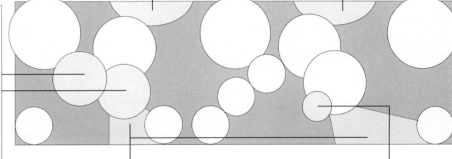

Rosa **'Graham Thomas'**. In recent years,
David Austin Roses has developed a collection
of old-fashioned roses, with all the disease
resistance of modern strains, which they call
their 'English roses'. This yellow shrub rose is
one of the best of this collection, and has one
of the longest flowering seasons. It is an
excellent companion for variegated *Aucuba*.
The colour combinations are subtle and refined.

Geranium **x** ***riversleaianum*** **'Russell
Prichard'.** This hardy geranium is a little gem.
It flowers from late spring through to the first
frosts without any attention. Its bright-pink
flowers are of a shade which combines well
with a range of yellows, which is why it has
been selected here.

Clematis macropetala. This small clematis
flowers in early to late spring with rich-blue,
nodding flowers. It is a good subject for a
container. In the wild, clematis grows up
through other plants, and here it has been used
in a similar fashion, allowed to trail across the
soil and grow up through plants naturally.

STEP FOUR
In-fill planting

Iris germanica. No scheme is complete without sword- or spiky-shaped leaves, and this is easily added by using a variety of iris. *Iris germanica* is the most commonly grown bearded iris and comes in a wide range of colours. Many are also scented. Any colour would be suitable here.

***Hosta sieboldiana* 'Frances Williams'.** Hostas prefer a moist, shady site, but many will grow in drier conditions and some can stand a degree of sun. The toughest are the *sieboldiana* varieties, of which this is one. The yellow-variegated leaves are a good companion for more finely divided leaves.

***Delphinium* 'Galahad'.** No garden is complete without a grouping of delphiniums. Their spectacular display in mid-summer gives impact to a border and provides an upright shape that contrasts with other plants.

Thalictrum aquilegiifolium (meadow rue). This stately plant has finely divided leaves and pink, cottonwool flowers, all of which combine well with other plants.

SOUTH-FACING LONG BORDER

Being in the sun for most of the day, this border offers hot and
dry conditions much loved by many of the more floriferous
plants. Here shape and texture have been maintained, but there
is more emphasis on the flowering plants.

The plants

Alchemilla mollis, 3½ in/9 cm
or 5½ in/1 litre pot

Artemisia 'Powis Castle', 5½ in/1 litre
or 6½ in/2 litre pot

2 Artemisia 'Silver Queen', 3½ in/9 cm
or 5½ in/1 litre pots

4 Bergenia cordifolia, 3½ in/9 cm
or 5½ in/1 litre pots

Carpentaria californica, 7½ in/3 litre pot
or larger

2 Ceanothus thyrsiflorus, 7½ in/3 litre pots

6 Delphinium 'Black Knight', 3½ in/9 cm
or 5½ in/1 litre pots

Fremontodendron californicum,
7½ in/3 litre pot

Geranium psilostemon, 3½ in/9 cm
or 5½ in/1 litre pot

2 Lavandula augustifolia, 5½ in/1 litre
or 6½ in/2 litre pots

3 Nepeta mussinii, 3½ in/9 cm
or 5½ in/1 litre pots

2 Osteospermum 'Stardust', 3½ in/9 cm
or 5½ in/1 litre pots

Phormium tenax purpureum 7½ in/3 litre pot

2 Rosmarinus officinalis,
7½ in/3 litre pots

3 Rosa 'Margaret Merril', 7½ in/3 litre pots
or bare-rooted

2 Salvia officinalis 'Purpurascens',
6½ in/2 litre or 7½ in/3 litre pots

2 Sedum spectabile, 3½ in/9 cm
or 5½ in/1 litre pots

Taxus baccata 'Fastigiata', 7½ in/3 litre pot

Vitis vinifera 'Brandt', 7½ in/3 litre pot

Quick tips

Bergenia cordifolia.

Flowering season: the main flowering season will be in late spring and early summer, although there will be colour throughout the summer.
Soil: any good garden soil. Improve drainage by adding sharp sand at planting time and incorporate well-rotted manure or compost around the roses in early spring.
Care: it is important to cut Artemisia 'Powis Castle' to within a few inches of the ground in mid-spring to prevent it from becoming leggy. The lavender should be trimmed over after flowering and again in spring. If the sage looks untidy and leggy in the spring, then cut it hard back to encourage re-growth from the base. Remove any dead wood from the roses in early spring.

During the summer season dead-head the roses and trim over the Nepeta to ensure a continuity of bloom. Stake the delphiniums with sturdy canes and tie the climbers into the wires as necessary.

Tools

Garden trowel • Spade
• Galvanised wire •Vine eyes

Planting plan

Main structural planting

Ceanothus thyrsiflorus (Californian lilac). The brilliant display of blue flowers always makes this plant the centre of attention in late spring, when it is at its best.

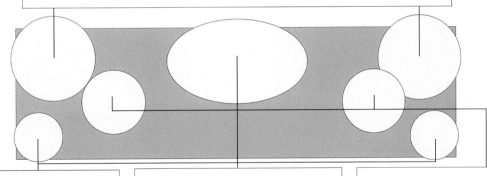

Lavandula angustifolia (Lavender). The common lavender flowers for a longer period than many of the hybrids, and the pale-lilac flowers provide better, more subtle support for the plant's neighbours.

Carpentaria californica. This evergreen shrub, festooned in white, saucer-shaped flowers in mid-summer, is the central feature of the border. It is certainly not a shy plant and will quickly establish itself.

Rosmarinus officinalis (rosemary). Of all the ornamental herbs, this is the queen. Its upright, spreading growth is unique and combines well in the cottage garden, as well as the warm, sunny border, where its blue flowers and aromatic, grey foliage provide solid support.

Artemisia 'Powis Castle'. The silver-grey, feathery leaves of this plant make it the perfect choice for any sunny border.

STEP TWO
In-fill planting

Phormium tenax purpureum (New Zealand flax). The bold, strap-like leaves make this a plant for impact. The purple foliage will combine and complement the blues, whites and pink of its neighbours.

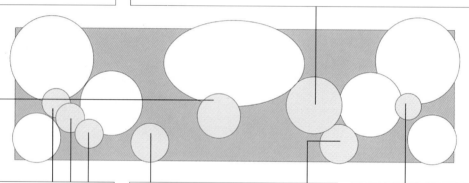

Rosa 'Margaret Merril'. For scent, this rose has no equal. The flesh-coloured blooms are held over glossy, healthy leaves all summer long.

Salvia officinalis 'Purpurascens' (purple sage). There are a number of ornamental sages available and each one is useful for combining with other grey-leafed plants, herbaceous perennials and roses.

Taxus baccata 'Fastigiata' (Irish yew). Upright, or fastigiate, conifers are a useful addition to a border when different shapes are required. All forms of the common yew are very tough plants, equally at home in deep shade as in full sun.

***Vitis vinifera* 'Brandt'.** This ornamental grape vine is an extremely useful garden plant. It will grow anywhere, including dry shade – although in these conditions the grapes can be forgotten. The foliage is attractively divided and colours up well in the autumn.

STEP THREE
In-fill planting

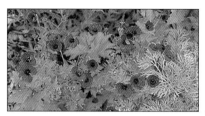

Geranium psilostemon. This stately plant is the tallest of the hardy geraniums. The bright-magenta flowers are held above self-supporting stems for a long period in mid-summer. Here the geranium works well with the light, feathery, silver foliage of *Artemisia* 'Powis Castle'.

Fremontodendron californicum (fremontia). Large, yellow flowers, carried on this evergreen plant throughout the summer, make this one of the special sights of summer. The combination of this plant with *Ceanothus* is quite breathtaking. *Warning: the dusty down on the leaves can irritate the skin, so wash your hands after handling it, or wear gloves.*

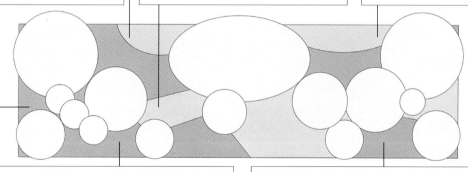

Bergenia cordifolia (elephant's ears) The large, leathery leaves of the *Bergenia* are used as a solid contrast to the more feathery foliage in the scheme – here the foliage of *Geranium psilostemon*. The bright-pink flowers are a welcome splash of colour at the end of the winter.

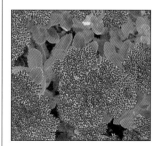

Sedum spectabile (ice plant). Late summer would not be the same without the flat heads of the *Sedum*. These bright-pink flowers attract butterflies and bees in great numbers.

Alchemilla mollis (lady's mantle). Every garden should incorporate this plant somewhere. The lime-green flowers combine well with any colour schemes and especially with roses and herbaceous plants. It will seed itself around freely and the seedlings seem to thrive wherever they appear. Here the *Nepeta* and lady's mantle look good planted as close neighbours. The contrasting shapes complement each other to create a pleasing combination.

STEP FOUR
In-fill planting

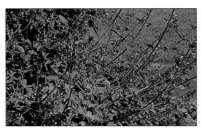

Nepeta mussinii (catmint). Pale-lilac flowers are held on long stems throughout the summer months and make a good companion for feathery-foliage plants like lady's mantle.

Osteospermum 'Stardust'. Mauve flowers, produced from spring through to autumn, keep interest going in the border throughout much of the year.

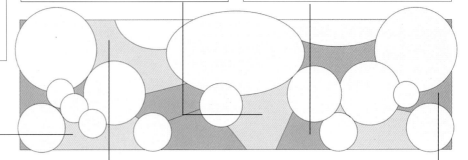

Delphinium '**Black Knight**'. The tall stems of the delphinium will pick up the fastigiate shape of the Irish yew and contrast with the other shapes in the plan.

Artemisia '**Silver Queen**'. This herbaceous *Artemisia* is more open than the shrubby 'Powis Castle' and dies down in the winter. The silvery foliage is finely cut, making a spectacular display.

WEST-FACING LONG BORDER

Warming up later in the day, the west-facing border offers one of the best, sheltered spots in the garden – a place where plants which would not survive in other parts of the garden may be grown. Some of the best effects in a garden can be created by using only a small number of different varieties. It may take a little courage to do, but the final show can be stunning.

The plants

2 *Choisya ternata*, 7½ in/3 litre pots
or larger

9 *Geranium* x *riversleaianum*
'Mavis Simpson', 5½ in/1 litre
or 6½ in/2 litre pots

2 *Lavandula augustifolia* 'Munstead',
6½ in/2 litre or 7½ in/3 litre pots

Rosa 'Nevada', 7½ in/3 litre pot
or bare-rooted

Rosa 'Scarlet Fire', 7½ in/3 litre pot
or bare-rooted

12 *Stachys olympica* 'Silver Carpet',
3½ in/9 cm or 5½ in/1 litre pots

6 *Yucca recurvifolia*, 7½ in/3 litre pots
or larger

Tools

Garden trowel • Spade

Quick tips

Rosa 'Nevada'.

Flowering season: the main flowering season is in mid- to late summer, a time when many gardens begin to look exhausted as they await the late-summer-flowering surge.
Soil: any good garden soil.

Care: the main task is to trim over the lavender after it has flowered and again in early or mid-spring. This will help to maintain the overall shape of the plant. Remove any flower heads that appear on the *Stachys* immediately. Remove any dead wood from the roses in early spring.

Planting plan

STEP ONE
Main structural planting

Rosa 'Nevada'. This large shrub rose is festooned in early and late summer with large, white, saucer-shaped flowers, making it the perfect choice for a large border. The size and colour of its flowers make it a good companion for *Rosa* 'Scarlet Fire'.

Rosa 'Scarlet Fire'. This large shrub rose has one spectacular display of brilliant-red flowers in mid-summer, followed by bright-red hips.

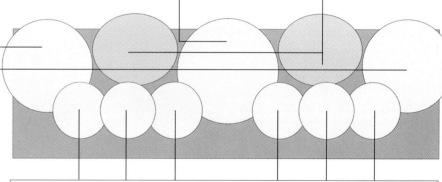

***Choisya ternata* (Mexican orange blossom).** One of the best evergreens available, it will grow in sun or shade and has a 'skirt' which comes down to the ground, smothering any weeds. The flowers in late spring have an orange scent and the leaves are aromatic when crushed.

Yucca recurvifolia. Spiky leaves add impact to a border. Here a yucca has been selected. This evergreen plant will develop tall spires of bell-shaped, white flowers in late summer, when the border will be at its best.

STEP TWO
In-fill planting

Lavandula augustifolia **'Munstead'.** This plant has dark-violet-blue flowers – much darker than the ordinary lavender. It is also lower-growing and more compact.

Stachys olympica **'Silver Carpet'** (lambs' tongues). The flowers of this plant are not particularly attractive and distort the overall shape. The variety 'Silver Carpet' is non-flowering. The combination of lavender, yucca and *Stachys* makes a good display on its own, without the addition of other plants.

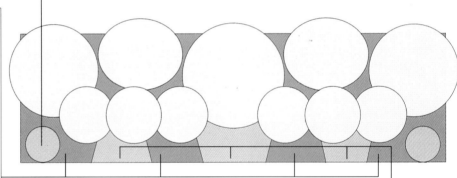

Geranium **x** *riversleaianum* **'Mavis Simpson'.** Flowers for most of the summer, combined with pale-grey-green foliage, making this a much-sought-after, hardy geranium. The geranium will grow through the *Stachys* in time, making it appear that the latter has pink flowers.

NORTH-FACING LONG BORDER

*This aspect is satisfying to plant up. Although lacking in bright
colour, this can be more than compensated for by using
the best foliage and scents.
Cool colours, such as blues, mauves, pale pinks and cream,
are the best choices for a shady site; hot colours, such as
reds and oranges, are much better in full sun.*

The plants

Amelanchier lamarkii, 7½ in/3 litre pot or larger

2 *Clematis* 'Perle D'Azur', 7½ in/5 litre pots

2 *Fatsia japonica*, 7½ in/3 litre pots or larger

9 *Hosta tokudama*, 3½ in/9 cm or 5½ in/1 litre pots

9 *Hosta undulata* var. *undulata*, 3½ in/9 cm or 5½ in/1 litre pots

2 *Jasminum humile revolutum*, 7½ in/3 litre pots

2 *Lonicera pileata*, 7½ in/3 litre pots

10 *Rodgersia pinnata* 'Superba', 5½in/1 litre or 6½ in/2 litre pots

Tools

Garden trowel • Spade

Quick tips

Hostas and *Rodgersia*.

Flowering season: this is less important with this scheme than with others, but the main flourish is in early to mid-summer.
Soil: any good garden soil. Add well-rotted manure or compost in early spring.
Care: slug pellets around the hostas will reduce any possible damage by snails and slugs. It is essential to plant the clematis roots away from the centre of the host plant to ensure that the competition between the two plants is minimal.

Planting plan

Main structural planting

Jasminum humile revolutum (shrubby jasmine). Mention jasmine and most people think of a climber with scented, white flowers. There are, however, a number of medium to large shrubby jasmines, mainly with yellow flowers. This one flowers from late summer through to autumn.

Fatsia japonica (castor-oil plant). This large-leaved evergreen looks more like an atrium plant than a garden plant, with large, glossy leaves held in palmate clusters. It has the advantage of being happy in quite deep shade, including that cast by a large tree.

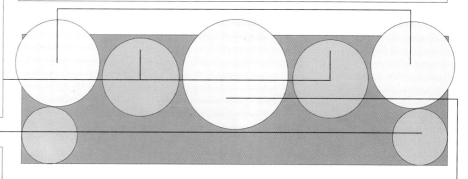

Amelanchier lamarkii (snowy mespilus). This has to be one of the choicest plants for any garden. It will grow in sun or shade, starting the season off in mid-spring with sweet-scented, white blossom and finishing the season with a brilliant foliage display in the autumn.

Lonicera pileata. This close relative of honeysuckle is an excellent ground-cover plant. Its outward- and upward- pointing branches offer an unusual shrub shape to the overall border.

***Clematis* 'Perle D'Azur'.** A good way of growing clematis is through and over other shrubs. Here a large-flowered, hybrid clematis grows through the shrubby jasmine, the blue flowers of the clematis contrasting with the yellow flowers of the jasmine in a particularly successful combination.

Main structural planting

***Rodgersia pinnata* 'Superba'.** This plant is generally regarded as a bog-garden plant, where it will quickly become established at the expense of other plants. In an ordinary border, so long as it has some shade and the root run does not become too dry, it is perfectly happy and will increase in size very slowly.

Hosta tokudama. This blue-leafed hosta is an ideal companion for the previous two plants, contrasting in both shape and colour.

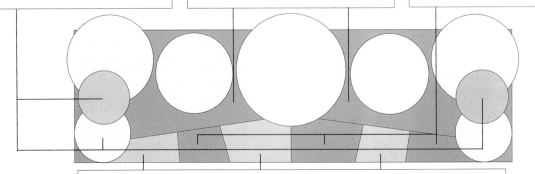

***Hosta undulata* var. *undulata*.** This is one of the finest of the small-leafed, variegated hostas. It is unusual, in that the variegation is in the form of a white line down the centre of each leaf. The lilac flowers in late summer are a bonus. Cool colours are ideal in a shady situation, and this combination of hostas and *Rodgersia* is perfect.

BOG GARDEN

*Areas of garden where the soil is constantly damp are a fairly
common occurrence close to a natural pond, lake or spring.
Many plants have evolved to cope with these conditions and, with
careful plant selection, the area can become an important
garden feature. If a natural area does not exist, it is possible to
create an artificial one, either close to a man-made pond or
perhaps in a border amongst other plants.*

The plants

Acer palmatum var. dissectum,
7½ in/3 litre pot

8 Astilbe x arendsii 'Venus', 5½ in/litre
or 6½ in/2 litre pots

5 Crocosmia 'Lucifer', 3½ in/9 cm
or 5½ in/1 litre pots

3 Houttuynia cordata 'Chameleon',
3½ in/9 cm or 5½ in/1 litre pots

5 Iris pseudoacorus 'Variegata', 3½ in/9 cm
or 5½ in/1 litre pots

4 Lobelia 'Queen Victoria', 3½ in/9 cm
or 5½ in/1 litre pots

Phormium tenax, 6½ in/2 litre
or 7½ in/3 litre pot

7 Polygonum bistorta 'Superbum', 3½ in/9 cm
or 5½ in/1 litre pots

Rheum palmatum, 6½ in/2 litre
or 8½ in/5 litre pot

7 Rodgersia aesculifolia, 3½ in/9 cm
or 5½ in/1 litre pots

4 Zantedeschia aethiopia, 6½ in/2 litre
or 7½ in/3 litre pots

Tools

Spade • Garden trowel

Quick tips

Astilbe x arendsii.

Flowering season: summer. The bold leaf shapes of the *Rheum* and the white-flowered *Rodgersia* make a dramatic combination. The hot colours in this border combine to catch the eye from every part of the garden.

Soil: any good garden soil, which is damp all year round, but not under water. If these conditions do not exist naturally, excavate the soil to a spade's depth and introduce a piece of butyl or plastic liner, perforated in several places to allow excess water to escape. Ensure that a hollow is created by mounding the soil around the edge, under the liner.

Care: provided the soil is constantly moist, these plants will grow in full sun. Apart from cutting down dead foliage in the autumn, there should be no maintenance required.

Planting plan

Main structural planting

Acer palmatum var. dissectum (Japanese maple). Often difficult to place, this low-growing shrub prefers a sheltered site and acid soil. The damp conditions in a bog garden simulate acid soil and the moisture ensures that any desiccation by the wind, so often the enemy of these plants, is minimised.

Zantedeschia aethiopia (arum lily). The creamy-white, funnel-shaped. flower spike of this plant is like the best porcelain and is a great favourite. Try growing it in a container.

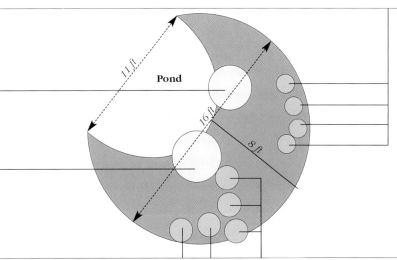

Pond

11 ft

16 ft

8 ft

Phormium tenax (New Zealand flax). Regarded as an exotic shrub, this plant is often thought to be a slightly tender sun-lover, but in fact it grows in marshy ground in its native home and is perfect for the water's edge.

Crocosmia 'Lucifer' (montbretia). The brilliant colour of this montbretia combines well with the hot colours in this scheme. Normally found in the flower border, this versatile plant is happy in a range of conditions, including those of a bog garden.

STEP TWO

Polygonum bistorta 'Superbum' (knotweed). Growing anywhere, these low, evergreen plants are useful for ground cover. The pink, bullrush-shaped flower heads are held over a long period in mid-summer.

Rheum palmatum (ornamental rhubarb). This statuesque plant resembles rhubarb only in the size and appearance of its leaves. It needs a moist soil, but can cope with drier conditions once established.

Pond

Astilbe x arendsii 'Venus' (false goat's beard). Feathery foliage and pink flowers with the texture of cottonwool make this a most desirable garden plant. It is often planted in the mixed border, where it dies out through lack of moisture.

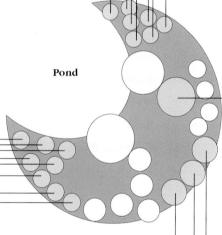

Houttuynia cordata 'Chameleon'. This ground-cover plant is at much at home in the flower border as it is in the bog garden. The multi-coloured, evergreen leaves are quite stunning.

STEP THREE

Iris pseudoacorus 'Variegata'. As has been said before in this book, the spiky shape is an important ingredient in any planting plan, and the bog garden is no exception. Iris is always a good choice. It will also grow in an ordinary mixed border and in shallow water.

Pond

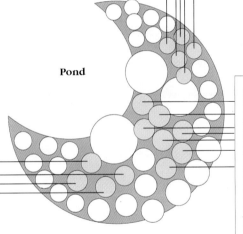

Rodgersia aesculifolia. The large leaves of this plant resemble those of the horse chestnut and make an excellent contrast for many other bog-garden plants. Although happier in a moist soil, it will grow in a shady border where the soil does not become too warm.

Lobelia 'Queen Victoria'. For brilliance of colour, no plant can outperform this tall-growing lobelia. It is short-lived and may need replacing every two or three years.

WOODLAND BED

Where there has been a lot of decaying matter in the soil over a number of years, usually in woodland, the pH of the soil drops and becomes more acid. Many plants have become dependent upon this type of soil, notably rhododendrons and most heathers, and cannot grow in alkaline conditions. This border has been designed to make full use of these acid-loving plants. The best site for it would be on the edge of woodland or in a clearing, where it can get some sun, but not too much, in partial shade.

The plants

Acer griseum, 8½ in/5 litre pot or larger

20 Hyacinthoides non-scripta bulbs

Kalmia latifolia, 7½ in/3 litre pot

3 Lilium martagon, 3½ in/9 cm
or 5½ in/1 litre pots or 3 bulbs

Magnolia x soulangeana 'Brizoni',
7½ in/3 litre pot or larger

2 Rhododendron 'Cecile', 7½ in/3 litre pots

3 Rhododendron 'Susan', 7½ in/3 litre pots

3 Skimmia laureola, 7½ in/3 litre pots

Tools

Spade • Garden trowel

Quick tips

Magnolia x. soulangeana and rhododendron.

Flowering season: spring and early summer. Combining magnolias and azaleas is the perfect way of ensuring that spring remains an unforgettable time of the year. The deep blue of the bluebell and the red of this rhododendron make a dramatic combination.

Care: the soil must have plenty of organic matter and peat so that it is rich and acid. The plants do best in partial shade, with some sunshine for part of the day.

Rhododendron 'Susan' with English bluebells.

Planting plan

Main structural planting

***Magnolia* x *soulangeana* 'Brizoni'.** A magnolia in full flower is one of those unforgettable sights of spring. Although not evergreen, the bare tree in winter has a fine, architectural shape which contributes to the structure of the garden.

Acer griseum (paper-bark maple). Many trees are grown for their bark, but nothing is quite like this unusual maple. The peeling bark is like flaked chocolate and is always a talking point. It will grow in any soil.

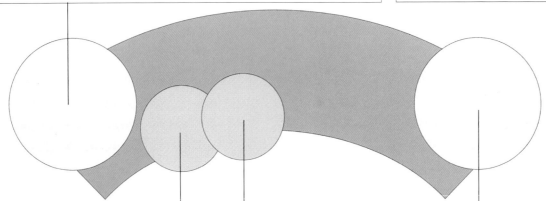

***Rhododendron* 'Cecile'** (azalea). The spring-flowering azaleas are a magnificent introduction to spring in any garden lucky enough to have an acid soil. They are good companions for magnolias.

STEP TWO

***Rhododendron* 'Susan'.** This evergreen
rhododendron, with its large trusses of red
flowers, will always dominate a border and it
has been used here for impact.

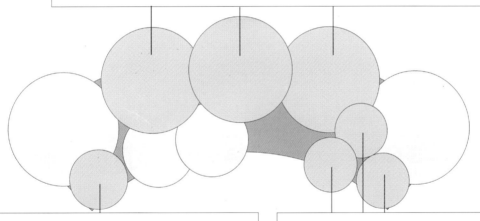

Kalmia latifolia (calico bush).
The early-summer flowering of this
evergreen shrub will help to
extend the flowering season of
this border.

***Skimmia laureola*.** In order to
get red berries on a *Skimmia*, a
male plant must be planted close
to a female. This variety is male,
so there will be no berries, but the
scent is better than that of the
female and will be most welcome
in early spring.

STEP THREE

Lilium martagon (Turk's-head lily). This unusual, bulbous plant will seed itself throughout the border, raising its head above the foliage where it is least expected without becoming a nuisance.

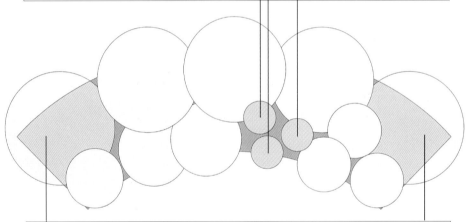

Hyacinthiodes non-scripta (bluebells). These are the picture-postcard bluebells of every bluebell wood and are the perfect companion for rhododendrons and azaleas.